How to Inhale the Universe without Wheezing

and other unconventional asthma lessons

Stephen Young

ISBN 978-1-4357-1510-3

Made in the United States of America

For my parents, Vernon and Betsy Young

Remember, O God, that my life is but a breath…

-Job 7:7 (New International Version)

Table of Contents

Introduction and Disclaimer-

The following story explores my experience with asthma. A very short summary might suggest I was afflicted and then cured.

It would be more accurate, I think, to say asthma slowly became more important in my life, until it became less important.

The process of writing the book increased my knowledge about my own asthma, however, this should not be taken as a recommendation for anyone else.

Asthma affects people in different manners and degree. Diagnosed with mild to moderate asthma, my experience varies greatly from someone who has severe asthma.

Much of this book describes me coming to terms with how little I really knew as I traveled the path from total ignorance about asthma to some consciousness about the condition. Asthma's impact on my every day life went from non-existent, to being the primary factor around which everything else was planned.

Then my relationship with asthma changed in many small ways and a few significant ways, but I wasn't always sure why those changes took place. Sometimes, just when I thought I knew, a new perception made old ideas obsolete.

I can say with some certainty that even if you have been diagnosed with mild to moderate asthma, attempting to reproduce the details of my experience in your own life would be disappointing, and possibly unsafe.

At the same time, if other asthma sufferers undertake an honest evaluation of their personal experience, they may be surprised by how much they don't know, and the benefits of learning more.

Section One

Unseen Spirit

Chapter 1

Known But Unnamed

In the dream, the attack wasn't entirely unpleasant – there was a sense of liberation as I sprinted across an open field on a bright day. But there was also the sense of dread, as each new breath became harder to draw.

My legs felt fresh though I'd been running for a long time. Even at a fast pace, I could add a sudden burst of speed at will. But unstoppable as my legs seemed, my lungs demanded more air. Each time I was ready to stop, thinking I need to catch my breath, a vague goal moved further into the distance. The pace kept up, as each breath grew tighter.

Then I was awake, trying to catch my breath, which cleverly evaded my clumsy grasp.

The clock was predictable. 3:30 a.m. About the same time it happened a few other nights in the past month.

The dream was over, but my lungs refused to calm down, gasping as if I really had been running. Actually, they felt as if the run continued at full power, though I was clearly laying in my bed in my very quiet suburban house. No matter how slowly I exhaled, trying to soothe the constricted airways, I simply couldn't inhale enough air.

My 10-year-old mind didn't have an accurate name for it. Commonly described as trying to breathe through a straw, or with a 50-pound weight on your chest, to me it was simply hay fever-related chest congestion.

Waiting it out would make it go away.

The summer was ending, fifth grade was about to begin. I roughly knew what asthma was, and that it had to do with breathing. In third grade, the music teacher instructed me to accompany my asthmatic friend Bill to the nurse's office after Bill seemed to strain for air during a spirited rendition of the folk song "Old Dan Tucker."

But no doctor or other grown-up had offered such a diagnosis, leaving me to assume my problem was something else. Or maybe, since my mom had Multiple Sclerosis that had been progressing in severity as I got older, I didn't want to think of myself as the sufferer of a chronic disease.

Hoping to return to dreams, I tried to lie quietly where the attack woke me. Sleep failed, no deep breathing possible.

Attempting to cough, it seemed maybe there was just some phlegm clogging up my airways. But the violent exhalation made inhaling even more difficult.

I tried moving to different positions, each for just a few minutes. Sitting up; leaning forward; kneeling beside the bed; returning to a reclining position on the bed; then rocking my head side to side; looking for some configuration that might allow a little more air to pass.

Sitting up seemed most conducive to good breathing, even though it was only the slightest of improvements. So up I sat, almost in a meditative position, without knowing what meditation was.

The constriction eased slightly at imperceptible intervals, but the breath still rasped lightly and I knew better than to lay down too soon. I took out a dusty *Mad Magazine* – already read at least five times – and sat flipping the pages. When drowsiness and relatively normal breathing returned, the sky's darkness was giving way to dawn.

Sleep came for an hour or two, but when I awoke, it seemed like only minutes.

Looking back, I did some things wrong, and a few things right. Fortunately, my ignorance was tempered by intuition (or perhaps it was the other way around).

It would have been reasonable to simply wake up my parents and ask for help, but at the time I thought of myself as big enough not to run to Mommy and Daddy right away. Since the attacks hadn't started during the day at that point, and my parents didn't see them as they took place, the conversation never really arose. The attacks were a nuisance, but I could deal with them.

Keep calm, try to relax, the breath will return like it did the other times. Getting through the early attacks, I learned that bad things happen, but usually they pass.

Years later, when I read the statement, "That which does not kill me only makes me stronger," from Friedrich Nietzsche, it made some sense to me.

Asthma can kill. Living with asthma takes strength, knowledge and patience. Acquiring that strength and knowledge and patience can positively impact other parts of your life. Asthma may have hurt my body temporarily, but the tools I unconsciously acquired to fight asthma may have strengthened me in another way.

Chapter 2

Emergency

The attacks generally came at night and had passed by the morning while I was in elementary school. Then, one summer while visiting my Grandparents' house in Rock Island, Ill., near the Mississippi River, the wheezing and the tight chest that started at night wasn't gone by morning.

My lungs continued to whistle and creak. My parents and grandparents didn't seem to notice at first, but my labored attempts at speaking made them see something was not right. Different opinions as to how to handle the situation were offered. Call a doctor. Get some fresh air outside. Take some antihistamine. Get some quiet rest.

The last two options were chosen more or less by committee. After sitting up and resting throughout the day, I began to feel better.

Other days that year I would notice the tight chest on fall mornings, particularly chilly ones, when the seventh grade gym class had to run half-a-mile outside to "warm up." I finished the course without walking, but still gasped for breath minutes after stopping.

It happened around the house sometimes - my parents and I labeled it as wheeziness, which we thought was a function of the allergies which had already been diagnosed.

One weekday evening after dinner, the wheezing started, but instead of reaching a plateau, it just got worse and worse. My breathing was clearly audible to anyone within a couple feet of me. I sat straight up in the easy chair by the TV, my fingers clenching the arms as I sucked for breath. My dad observed me in that pose for a while, consulted with my mom, then announced we were heading to the Emergency Room.

At the hospital, we waited for three hours. Sitting quietly, eventually finding the energy and interest to page through a month-old copy of *Time*, I regained my breath during the wait. But as the feeling of normalcy returned, the idea that the doctor wouldn't see the symptoms and would think this was some kind of scam on my part seemed plausible.

When the doctor finally saw me, he pressed the stethoscope on my back and asked me to breathe. He must have heard the smallest air sacs still whistling with obstructions.

He said asthma was the problem. It could be treated with medication. He said some people suffer for a while, then grow out of it. My case didn't seem too bad — it wouldn't kill me, he reported cheerfully. He said not to take antihistamines, but instead to fill the prescription he was about to write for me. The pad said "Marax."

It was sort of a Buddhist lesson — the recognition that life features discomfort — discomfort that sometimes doesn't just disappear on its own. However, the script he wrote also illustrated that there is a way to confront that discomfort.

Section Two

Treatable Illness

Chapter 3

Drugs

24-hour drug stores did not exist in the far south suburbs of Chicago in the early 1980s, so the prescription waited.

The next day, Dad brought home the pills: pale blue trapezoids with rounded edges about the size of a typical aspirin. The shape, with slanting, angular edges, suggested the future to me.

The scribble on the doctor's pad had been translated onto label of the opaque brown prescription bottle with the child-proof top. The futuristic pills inside were also inscribed: "MARAX."

I was told to take the pills for tightness in my chest or outright wheezing. The doctor said it might have a stimulant effect. I carried one in my pocket with me to junior high and told my friends it might make me a little jumpy. We all thought that was funny.

The asthma remained calm for a few days after visiting the doctor. Then, I woke in the middle of the night, after dreaming of running and struggling to pull in breath. I left my bed and went to the medicine cabinet where the pills had been stored.

The relief it offered seemed so inviting, but something held me back from taking it. I think by that time in my life I had read both *Go Ask Alice*, the supposed diary of a drug-abusing teen who goes from light experimentation to full-blown addiction-inspired prostitution , as well as Kurt Vonnegut's *Slapstick*, in which the main character becomes addicted to "tri-benzo-Deportamil" after just one try.

The anonymous author of *Alice*, with her sudden, but seemingly credible, spiral out of control, and the stark poignancy of Vonnegut's description were suddenly prominent in my mind.

I stared a long while at the pill. I'd never had any second thoughts about prescribed medicine before, but, in the middle of the night, wheezing with increased intensity, this seemed different. It wasn't a course of antibiotics that starts and finishes at clear points. This was going to be the way I reacted to my body – by relying on something outside of myself.

And then, after several moments of thoughtful discomfort, it didn't seem to matter that much. I took the pill and went back to bed.

Waiting for the jitters to come, I remember feeling only vaguely better before falling asleep.

From then on, pills were taken without much thought. Abusing them in any way did not cross my mind– they didn't make me feel that great in and of themselves. But when the need arose, I took one.

The pills were tools to cope with my problem, and I recognized that an optional tool is better than no tool at all. Taking the pill was a choice, I didn't have to. The responsibility was mine. Though I wouldn't read the quote for years later, it was something like what Epictetus must have felt when he wrote, "God has entrusted me with myself."

It may have been only one step beyond ignorance, or it may have been the key to the whole problem, but either way, one step is better than no steps.

Chapter 4

Some Control

I'm not sure how much the pale blue trapezoids helped. After swallowing one, it would be half an hour, maybe longer before anything detectable happened. The pills didn't really offer complete relief either, but they seemed to calm my system down enough for the breath to move more naturally.

Perhaps it was only psychosomatic, since the doctor suggested it might happen, but I did feel a little nervous after taking the pill. Something of a foot-tapper in school, always trying to expend a little nervous energy, the pills either made me more conscious of this, or fed the foot-tapping.

The pills were used seasonally – always in the late summer when the ragweed blossomed - every few days for a month when the pollen was heaviest, and they worked. If wheezing was audible, or if I just felt uncomfortable, the pill would be taken before sitting quietly and waiting for the lung-relaxing effect.

A sensation of tickling inside my ribcage could be felt, which seemed to accompany the relief, or sometimes the onset of an attack.

I didn't think of myself as an asthmatic so much as someone who had a little asthma which was easily handled with medicine. As snow fell and the autumn faded, I virtually forgot about the condition.

In my adolescent mind, this one apparently effective tool, recommended to me by a professional Medical Doctor, was the best modern medicine could do - and therefore asthma wasn't really a problem. I tried to ignore the first diminished inhalations, but when they couldn't be ignored, I took the little blue pill to ignore them again. Push down the symptoms and get on with it.

The pattern continued for a few years. Instead of thinking about what happened when my lungs tightened up, I had a shortcut to stop thinking about the fact that my lungs sometimes tightened up.

I thought that cycle could continue indefinitely, but my ignorance was about to be exposed.

Any tool, even a little bit of information, can be a wonderful thing to help you take control of your world, but it can lead to dangerous complacency.

Many religions and philosophies advise adherents to seek knowledge. Sometimes, shallow knowledge is forcibly replaced by deeper knowledge, even when it is not sought.

Chapter 5

The Drugs Don't Work Anymore

I settled into a predictable routine with Marax. Discomfort had not been eradicated. Sometimes I forgot to carry the medicine with me, not noticing until an attack had started. The attack would be handled the old way: trying to stay calm while feeling it grow worse, plateau, and then slowly recede.

After one of those attacks, I always made sure my parents knew the medicine was gone.

The pills seemed to work through junior high. I had the attacks, I took the pills, my lungs loosened a little, and soon my lungs were forgotten again.

Still, on the worst August nights, when the weeds were in full bloom, I'd have trouble sleeping (either from the asthma or the Marax, which did indeed include a fairly potent mix of stimulants).

Lying awake, waiting for calm and sleep in my upstairs bedroom, I moved around the house after everyone else was asleep, taking a sheet down to the couch in the living room and waiting there, hoping the air one-story lower would be more conducive to breathing. Eventually sleep would come, most nights. Then, one early morning in late August, just after the ragweed started releasing its pollen, sleep was impossible. Why not use those weird quiet hours to do something?

At that age, being a writer seemed like a decent vocation, but I didn't really write much, except when forced at school. Awake with nothing to do, why not start writing?

I liked science fiction and had an idea in my head for months, possibly years (probably very similar to the initial idea of many would-be science fiction writers) about galactic explorers who set out to reach the end of the universe. It seemed like a long way, but all night is a pretty long time too.

Filling the pages of a spiral notebook for hours, the story ended as the sun rose. Feeling more tired than if I had just tried to stay in bed, but also somewhat giddy with a sense of accomplishment, I went to bed for a couple hours.

It would be years before I ever thought of asthma as anything but a liability, but now I can see that discomfort sometimes brings inspiration. Some inspiration may never have come without the discomfort.

Marcel Proust, whose *In Search of Lost Time* is regarded as one of the great works of 20th century literature, finally sat down to write his masterpiece after his body was racked with sickness (including asthma). In the final volume of the largely autobiographical work he describes the process through which he finally "captures time" by carefully recalling the past, and notes, "[I]t is only when we are suffering that we see certain things that are hidden from us…"

The story I wrote that night certainly wouldn't be mistaken for Proust. I don't think too many people ever read it other than my parents. But who knows how many years it would have taken for me to write something on my own without my lungs forcing me to stay awake.

Chapter 6

Exercise –Forced Heavy Breathing

At the urging of my Dad, I had joined the cross country team in seventh grade, before being diagnosed with asthma. I liked the idea of being able to run a mile and half race – like those old dreams that used to wake me up, it seemed at least partially free and fun.

But in my 12-year-old reality, those great distances were not necessarily enjoyable to run, and I wasn't terribly good at running them. The aches and pains of conditioning my body weren't too much fun either.

Sometimes breathing was a little trouble during practice, but I seemed to get through to the end.

The first minute of running felt good, and then it became a little harder to breathe. After about 5 minutes, my leg muscles began to burn, and eventually my ankles ached. I forgot about the breathing and tried to estimate the number of steps needed to finish that particular exercise or race.

Now officially diagnosed with asthma, I saw an opportunity to give up on cross country in the eighth grade. At the beginning of that school year, one of the coaches said they were looking forward to seeing me on the team again.

"Oh, well," I said, "after being diagnosed with asthma recently, I'm not sure if I should." (The doctor never said one way or another whether I should be running, but it seemed to make sense to me.)

"Actually," the coach – a veteran science teacher who always looked fit even as she approached retirement age – said, "I have asthma too. One reason I run is to help break up the congestion."

"Uh, really," I said.

Yes, and she went on to tell me that she thought it was important to exercise to keep one's lungs in shape.

So, for that year, I stayed on cross country. And, I saw there was some truth in what she told me. Exercising right in the middle of an attack didn't seem like a good idea, and some days the wheezing did appear to cut into my ability to exercise. It did seem running might have made me cough at the start, but when the run was over, I always felt a little better than before I started.

Forcing my lungs to work a little harder was a little uncomfortable at first, but when it worked correctly, the lungs open up a little bit.

Aside from learning about how exercise can help breathing, I also learned that knowledge is not always where you expect it. Yes, junior high teachers have knowledge, but to me at the time, that knowledge hardly seemed practical. Looking back now, learning about exercise and asthma might have been as important as anything else I learned in eighth grade.

Chapter 7

Hell in a Church

It was supposed to be a spiritual experience, and now I understand it was. But at the time, it just seemed unpleasant.

The church youth group I belonged to occasionally went on overnight retreats. In the tenth grade, our south suburban church headed into the north part of Chicago (Rogers Park specifically) to spend the evening at a church there.

Church retreats didn't seem as fun as hanging out with my non-church friends and playing poker, but the retreats weren't terrible either. Except at this one, I forgot the bottle of little blue pills.

Without an attack since the previous fall, still in the mindset of mid-summer, I didn't think of it too much. But, as we walked the city streets absorbing the culture, I felt the clenching in my chest. I realized the Marax was 35 miles away. Trying not to make a big deal about it and hoping it would pass was the only strategy I could imagine.

It was almost dinner time. We headed into a rib joint, which seemed like a good idea. I got rib tips, the cheaper, fattier pieces on which meat clings to a bit of bone or tendon. The spicy sauce stung my tongue, but I wolfed down the greasy meat along with a piece of white bread and French fries that were more brown than golden.

The food tasted great, but my airways seemed to swell. Wheezing audibly, the walk back to the church was uncomfortable. One of the youth leaders eventually asked if everything was OK. No big deal, I've got a little asthma, and I forgot my medication, but it shouldn't be a big problem.

He said we could stop at a drugstore, and we did. I'd never really looked at the over-the-counter options available for asthma, though I had seen the Primatene Mist ads on TV hundreds of time – where the mist works as fast as the sprinter on a 50 yard dash.

I read the packages of both the mist and the pills, but felt uneasy about taking anything without doctor approval (and I didn't want to spend money I didn't have to spend). I bought a candy bar, but no medicine.

The first activity lined up at the church was gym time – which was novel, since our tiny suburban church on the edge of the cornfields wasn't anywhere large enough to have a gym. Trying to play basketball made me feel even worse. I tried to rest a little, before being enticed into playing Pig, which didn't involve running.

After quickly earning all three letters, I gave up and went back to just watching while sitting against a wall. I just wanted to maintain, not get worse. Slowly, perhaps every five hundredth breath, slight relief could be perceived. Finally, after two hours, I felt not quite normal, but beyond the crisis.

Sitting and staying awake was the best thing to do as the games calmed down. There were other rooms for sleeping, but not being able to rest, I headed back to a library where the minister of my church was working on a sermon for the next morning. I asked him about how he wrote them and whether he ever recycled them. It was an interesting conversation, but I finally went to bed, at about 4 a.m.

While I didn't learn either lesson fully at the time, there were clear messages about being prepared, and finding knowledge while others are sleeping.

Chapter 8

Trying to Ignore Reality

After the church trip, the Marax stayed with me during asthma season. Yet, the once predictable pills weren't working as well.

One late summer night right before high school started, the Marax didn't seem to do anything. I sat in my room waiting for that first sensation of relation inside my chest, but it never came. The wheezing just did not stop.

I walked down the stairs in my house to where my parents were watching television. It was exhausting, making me breathe even heavier. They asked me if I had taken my medicine. Yep, I said.

Sitting with them, watching an old movie on TV (*Heaven Can Wait*, the original with James Mason, not the remake with Warren Beatty), I gripped one of the arms of the chairs tightly, trying to will my lungs to accept a little fresh air.

After a little while of listening to my audible breath and seeing me strain for air, Dad decided another trip to the emergency room was in order. On the way, I assumed I'd sit and wait while the attack went away, then the doctor would say, "See, the medicine worked. Why did you bother me?"

But it was different. I didn't have to wait all that long. Then the doctor walked in and had just said hello, when he motioned to his stethoscope and said, "I hardly need this." He could hear me breathing.

After the examination the doctor said I wouldn't need to refill the Marax prescription, that there was another drug that was much more effective, delivered in mist.

He pulled out a small metal cylinder a little bigger than my thumb, which was housed in a plastic tube molded to the shape of a periscope.

The inhaler contained medicine - he called it by its generic name Albuterol. The doctor showed me how to exhale, press the small silver canister and then inhale the puff of medicine the device released. It made a light whoosh sound.

I took one puff, but I think it got stuck in my throat. I was told I should take two, and try and keep the inside of my mouth open.

The second one got through. It was like a miracle. Actually, at the time, it was a miracle. After a few short seconds, the relief started to kick in. My chest relaxed, finally able to expel the air trapped in my lungs and able to absorb fresh air.

After about a minute, I hacked up a rather nasty chunk of phlegm and felt even better. That was the problem, it always felt like – there was gunk in my lungs, but no matter how hard I tried to cough it out, it seemed only to get lodged more deeply.

A sullen teenager who wasn't prone to public displays of emotion, I must have expressed uncharacteristic appreciation to the doctor.

Chapter 9

Heaven in Spray Form

Talk about better living through chemistry. Modern medical science had shown its God-like power to me.

Not that I wasn't a believer before, but it's hard to overstate what a revelation the Albuterol puffer was in my life.

It represented relief, it represented certainty, it represented freedom. If the asthma attacked, I had a clear defense.

In his memoir *Self-Consciousness*, John Updike describes using an asthma inhaler as being akin to breathing in blue sky. That's close, but I think the feeling is actually better.

Compared to all the years of perceiving an attack, and then waiting for slow, incremental relief, this really was like a silver bullet. It was almost like the asthma was ignorable.

In spite of the near miraculous effect of the drug, I had to remember little things, such as bringing the device with me when it might be needed. It also took time getting used to taking medication in public.

Sneaking away some place by myself seemed appropriate. I wasn't ashamed of the illness, but to my young midwestern mind, taking medication in public would draw unwanted attention to a private issue – perhaps even be perceived as a cheap plea for sympathy.

At that time, in my experience, you didn't see a lot of people simply whipping out their inhalers in public. While it's possible to perform the operation discreetly, it's not nearly as discreet as taking a pill. Beside the inhaler itself, there's that little whisp of sound when the medicine is pushed out of the canister.

Sometimes, I learned, there wasn't much choice. Stuck in a classroom or some other place where it's tough to simply walk away unnoticed without an excuse. I took the puff regardless of where I was.

Young people would simply stare or ask what was going on. Sometimes my fellow teenagers jokingly asked if it was a good buzz.

Adults tried to simply appear sympathetic, or mention a friend or loved one who also used an inhaler. I usually responded enthusiastically about my medical experience.

Starting work as a caddy at a private golf club, I seemed to need the inhaler after about 14 holes in the late spring and early fall. This also led to conversations about asthma, but no revelation on my part that the attacks were so regular. Because I could make them go away quickly, the asthma occupied virtually no space in my conscious mind.

This was OK, as it allowed me to keep caddying and which made me eligible for a college scholarship. My life would have never followed that path if I had to wait for a pill to work while I recovered my breath on the 15th tee.

Though I would later have mixed feelings about Albuterol, my life would be much different today without it. Technology may be a double-edged sword, but it can be blessing from God.

Section Three

Stupid Inconvenience

Chapter 10

Lessons I Didn't Know I Was Learning in College

After high school, I attended Northwestern University. While there, I used the inhaler with some regularity during the fall. After living in a house with non-smokers for all of my life, I am exposed to tobacco smoke. I tried to be tolerant, but sometimes it bothers me. So I used the inhaler.

When I start wheezing on the way to class, I used the inhaler.

I stayed up late, drank alcohol and sometimes ingested other mind-altering substances, always with the inhaler at my side.

Sometimes I used the inhaler preventatively if was going someplace where I might encounter a cat or dander from other animals.

There were times during the four years of college when the inhaler must have been reduced to nothing but a mist of propellant containing only the barest hint of medicine. But I kept on using the inhaler, taking comfort from it. When the mist finally stopped altogether, I prepared to call the doctor for a refill, but I didn't let the lack of medicine slow anything down.

I knew what to do, and had total faith that the little silver canister would get me through. And it did, even when I unwisely played intramural floor hockey in the midst of asthma season, running to the sideline for the inhaler if it got too intense.

Difficult to see until later, it's now clear that I had some knowledge, some tools and some judgment. Using them as well as possible at the time, perhaps my over-reliance and unwillingness to cut back on anything in order to relieve my asthma showed I wasn't just using my knowledge and tools; I might have been abusing them.

I had a fair amount of judgment, but I didn't always exercise it wisely.

Fortunately, through the grace of God, I suppose, the experience didn't lead to any trips to the emergency room – though there was another kind of trip to the hospital. I went totally by choice, but it was still probably ill-advised and should have been more thoughtfully considered.

Chapter 11

Rejected by Science

The classified ad seeking asthma sufferers for a scientific study was published in the student newspaper just as spring semester wrapped up. Finally, I thought, a chance to literally profit from a liability.

Advancing science and making a little cash simultaneously seemed like a win-win situation.

After sailing through the pre-screening on the phone, a face-to-face appointment was scheduled.

When the day came, I took the El from Evanston to a hospital on the north side of Chicago. After signing some forms, which essentially absolved the hospital of all liability if anything went wrong, the doctor arrived. He explained what would happen.

They were testing a new medicine for people with mild to moderate asthma. I would take a physical examination and have my asthma tested to see if I was appropriate for the test.

After the basic exam, a nurse repeatedly poked my skin with a sort of needle and then smeared common allergens on the tiny wounds. The ragweed test popped up like a small volcano on my arm, while a few others swelled slightly. The nurse and doctor both appeared impressed with the volcanic reaction.

The next part of the test involved me breathing the allergen that bothers me to see how I would react. Since my skin reaction was most violent on ragweed, that's what I got to inhale. A small plastic mask which introduced the ragweed to my lungs with ruthless efficiency was fitted over the bottom half of my face.

The nurse measured reactions over various intervals of time – first every ten minutes or so, and then out to 20 minutes. I was starting to feel a little bad by the first test, which consisted of blowing as hard as I could into a machine, but the machine didn't measure much objective difference. The next two tests measured my lung capacity as a little worse, as expected, but the doctor said to take a break for lunch, and come back for the next test in an hour.

The exact meal has been forgotten except for a side dish of cut corn (though that could be a manufactured memory based on later experiences), but about halfway through, the bad feeling intensified. Not much stopped me from eating, so even though I was uncomfortable, I continued to pack it away.

After finishing, I usually would have reached for my inhaler, but then realized, I couldn't. As part of the test, I wasn't supposed to use any medicine until the test was complete.

Pressure on my lungs grew as I made my way back to the doctor's office. Even moving slowly was uncomfortable, but the elevator wall provided a place to rest.

My breath might have been heard before I entered the office. Even before hooking me up to the machine that tests breathing strength, the nurse looked a bit concerned. When the test started, my best effort could barely move the instrument.

"That's quite a reaction," the doctor said. They had me sit down and conferred for a while.

My asthma was too serious for this study – but since I spent the day there, partial compensation was available. Then they offered me some rescue medicine, which I gladly accepted. They had me wait a bit to observe me. When it was time to go, they insisted I take the medicine with me.

Getting home wasn't a problem, but the unpleasant feeling of being overwhelmed with allergens remained for a couple days. When the $150 check arrived in the mail weeks later, my mind (and body) had a sort of flashback to that horrible feeling of the test, leading to the

realization that no amount of money is worth feeling so
bad.

Chapter 12

Unexpected Relief an Ocean Away

After graduating from college, I decided to take advantage of a travel program for students and go to England for six months.

The objective had nothing to do with getting away from asthma. In fact, no plans were made to accommodate my asthma, other than bringing the Albuterol inhaler along.

This wasn't a problem. I left the U.S. in mid September, right when the ragweed is most potent. Arriving in England, it took a couple days to adjust, and I got sick with a cough, but soon felt fine.

After the cough passed, I didn't really notice the asthma for the rest of the visit.

Not careful about avoiding the things that bother me, I settle in to work as a bartender in a pub. At that time, in the early 1990s, English pubs were incredibly smoky places. The smoke didn't bother me.

Cash tips were rare, but many customers offered to buy drinks for bartenders. This can lead to a fair amount of drinking on the part of bartenders, night after night. Didn't bother me.

I can't say that asthma was completely forgotten, but it slowly took less precedence in my mind. I became casual with my inhaler, sometimes not toting it along when going out. It wasn't a problem.

Perhaps I wasn't consciously aware of it, but breathing relief in England showed that little things can make a big difference.

Of course, going to another country isn't exactly a little thing, but what was different? Was there less pollen? Something in the food, the water?

Beyond the wonder of the positive change was the message that things don't always get worse. I read some Zen books in London that probably conveyed that same message, but failed to make any sort of connection, until reflecting on it now.

Sometimes a health surprise isn't a bad thing and sometimes you find relief when you aren't really looking for it.

Chapter 13

Home Insecurity

I returned to the U.S. in mid-April and continued to feel fairly free of asthma symptoms for a few weeks.

Then I went to see a movie with my sister. The movie was *Flatliners*, a moody medical thriller with Julia Roberts. I was enjoying the cheesiness of the movie and the popcorn (something that wasn't usually offered in British movie theatres) when I felt the pressure coming on.

Damn. I hadn't thought about asthma for months, and here it was. I dug around in my pocket for the inhaler despite my certainty it wasn't there.

Trying to stay calm, my breath was becoming more and more constricted with at least a third of the movie left. Sometimes just sitting still helped, so still I tried to be.

I made it through the movie and back home. The attack was already diminishing by the time the inhaler was located at home. After that episode, Albuterol became, like keys or a wallet, something that had to be in my pocket before I could comfortably leave the house.

It didn't seem to be a problem until my coverage ran out under my parents' health insurance plan. Working part time at a catering service while trying to write screenplays, medical benefits were available for neither.

Lacking excess money, visits to the doctor stopped. The plan was to replace doctor visits and prescriptions with non-prescription medication. Recalling the asthma attack in the church, I regretted not taking advantage of such a convenience. Primatene mist and pills were purchased.

It seemed like a bold act of independence, until the stuff is actually ingested.

Unlike the generally neutral (or, to my taste, even lightly sweet) flavor of the Albuterol, this was harsh: kind of sickly chemical sensation on contact that quickly turns bitter. Each time I took the medicine, I couldn't help rubbing my tongue roughly on the roof of my mouth and then sticking it out with squinting eyes and puckered lips.

It didn't work as well either. The non-prescription mist offered some relief, but it wasn't as immediate or complete as the Albuterol. The non-prescription pills made me a little jittery and reminded me of Marax. I stayed away from the pills, but the Primatene inhaler seemed like the cheapest and best way to go.

I usually used the non-prescription brand, but saved an old canister of Albuterol, with just the tiniest bit of vapor in it, and sometimes if the Primatene wasn't working, a hit was taken from my nearly exhausted old friend, which seemed to help.

The asthma started to get a little worse.

I don't know what was learned at this time, but looking back, it's clear I didn't learn the previous lesson about money not being worth the sacrifice of asthma relief. It was a lesson that was going to hit harder.

Section Four

Growing Burden

Chapter 14

Flirting With Alternatives

The asthma finally seemed to stabilize after getting a little bit worse with the non-prescription medicine.

I talked a little with other people who had asthma. Just trying to make conversation, I heard many interesting things. For example, my friend Dan, who drank two large glasses of water when he felt an attack coming on.

During the summer, the non-prescription medication, with some exercise and the water trick, kept everything under control.

Taking a vacation to the Canadian Rockies with my girlfriend in the late summer, it was all good fun until I woke up wheezing as we camped out near the border. I tried to use the non-prescription inhaler, but it didn't really help. Fortunately, I brought the Albuterol, which still made a slight whoosh sound when the canister was pressed. It helps. We stayed in a hotel the next night, but hiking around the clear blue water of Lake Louise near Banff the morning after a rare August snow has fallen, my lungs rebeled.

Usually, doing something physical helped to clear the problem a little bit, but the more I walked, particularly on inclines, the more my lungs constricted as if they were squeezed by a python.

The medicine helped a little, but my need to stop and rest frequently cut our hike short.

For the rest of the trip, the breathing problems came and went, but by the time we got home, the appeal of self-treatment had worn thin.

I made a doctor's appointment and prepared myself to pay for it and the medication.

The doctor asked about medication, and I mentioned the non-prescription stuff. He didn't say anything in response, but he seemed to offer a disapproving look.

At the end of the visit, he wrote a prescription for Albuterol, my true goal of the visit. The doctor asked what else I was doing to maintain my asthma. "Uhhh..." I dumbly replied.

He said there were newer treatments and medications available that helped to reduce attacks, not just symptoms. Management was the key, he said, and that included seeing a doctor regularly.

I said I'd be back, but after the prescription was filled, I didn't have a second thought about it until now. I should have learned that when your body is trying to tell you something, listen. Instead, I misread that lesson, incorrectly inferring that the prescription Albuterol inhaler could get me through anything.

Chapter 15

More Heavy Breathing at the Movies

The Albuterol puffer is almost completely used before a return to the doctor crosses my mind. Frequently, the puffer is pulled out and used halfway through a movie at the local theater, where I get free passes, a rare perk of writing arts reviews for a small-town newspaper.

The movies don't usually have much to say about asthma, but one scene sticks in my mind.

Rush featured Jennifer Jason Leigh as an undercover cop and Greg Allman as a nasty drug kingpin.

I don't remember much about the movie except a short scene that was fairly insignificant in the context of the whole movie.

In the scene, the undercover cop shows up at a drug dealer's apartment. The gatekeeper whom she addresses, a heavyset, seemingly sedentary man, sits on a couch and takes a hit off what appears to be a Primatene mist inhaler – not unlike the non-prescription puffers I had used recently. The film's sound people got the whoosh just right.

The first puff is interesting in itself, because usually the presence of an inhaler in a movie signifies nerdy comedy or medical tragedy. This was something else – just simple maintenance. Or was it?

As the scene progresses, the man takes another hit off the inhaler. And another. And another.

The insides of my lungs kind of crawled at the thought of those tickling little fingers fluttering inside the air passages after just one or two doses of such medicine.

But taking 6 doses within a couple minutes – I could imagine the sweet taste in the mouth passing into bitterness, before another makes the tastes even more intense, and the fluttering fingers reaching way down into the pit of my stomach.

Beyond the feeling of revulsion the scene inspired, it also made me think about what we consider drug abuse and the circumstances that foster it.

I know it was only a movie, but it made me think about what would happen without a choice but the non-prescription harshness. What if that seemed like the only available relief; what if I overused, gaining tolerance and diminished effect; what if I didn't know what else to do?

Such a vicious circle would be so hard on the body, but would I listen to my body? And would listening and reacting constitute addiction? As usual, about halfway through the movie, after I finished a bucket of popcorn, I pulled out the Albuterol inhaler and sprayed another blast into my lungs.

Chapter 16

Some Moments of Bliss

In the context of this book, it probably seems as if my life is focused around asthma, but other things happened too. After a few years of working at a small town newspaper and earning a Masters degree, my girlfriend who traveled to the Canadian Rockies with me becomes my wife. We move to the northwest suburbs of Chicago where she already has a good job.

Little thought was given to how married life might impact my asthma, and at first I don't have to think about it. Switching between the non-prescription inhalers and the prescription inhaler seems to keep everything under control.

For a while, neither gets much use. Meanwhile, life without health insurance had become familiar, though not comforting. Being added to a plan seems appealing, and indeed, I find myself eligible for my wife's plan.

After several months of working as a freelance reporter at a local newspaper, I was offered a full-time job.

Suddenly, after going for a few years without health insurance, I found myself with two health plans.

And yet, I don't go to see any doctors for some time after the benefits kick in, still relying on the aging inhalers obtained months earlier.

Perhaps if more serious problems had appeared at the time, a doctor's appointment would have been scheduled. But, coasting seemed suitable, and it was easy...

Sometime in this period, I should have learned another lesson, like to take advantage of low-cost medical care when it's available. That didn't seem important though.

Just enjoying married life at the time was my focus, and that likely beneficial. I didn't know it at the time, but the lesson I learned had to do with settling into life as it comes and making the most of it when you can.

If you've got spontaneity and joy in your life at a particular moment, revel in it; it's permissible not to dwell on your problems and their implications all the time.

Even a good, single, unmedicated breath can serve as pleasurable refreshment, so enjoy it. And that's what I did, for a while.

Chapter 17

Slow Descent

The problems were minimal for some months – symptoms were only felt during fall asthma season - but the second fall asthma season after marriage was brutal.

I found myself with health insurance again, but still didn't make an appointment, until it got really bad.

The doctor was nice and explained that asthma sufferers should be taking inhaled steroids, preventative drugs designed to keep attacks from happening, instead of just taking rescue drugs, which were supposed to stop attacks once they already started.

I filled the prescription for both the steroid and Albuterol. I knew enough not to mistake the inhaled steroids for the type of steroids abused by athletes and body builders, so I was just about ready to embrace the medicine. But that first puff brought a grimace as the mist coated my airways. A terrible bitter taste spread through my respiratory system. Ugghh.

After two days of not being able to rinse the taste out of my mouth, I went back to relying on the Albuterol. That was fine, until that prescription had to be refilled a few months later, which required a doctor's visit.

Asked by the doctor if I was using the steroid, I hesitated before answering honestly that I had not. The doctor (who was different from the doctor at the previous visit) said I should take the steroids, but that there was another drug that might help as well.

Serevent was also supposed to act more as a preventative than a rescuer. The first puff out of the green plastic holder was more like a blast. Boom!

My airways seemed to dilate with a sudden shock - almost violently. I felt the need to shake it off, as if I just downed a straight shot of whiskey.

Man, were my airways clear though. I thought I was on to something.

That steroid box was never opened. But the Serevent was used religiously, even to the point of cutting back on Albuterol.

Over the next year, each time I went to the doctor, the Serevent and Albuterol prescriptions were refilled, while the doctor (whoever was available at the HMO, there seemed to be so many) usually said inhaled steroids would help, but to do what works for me.

I thought little of it, until the day the Serevent didn't really give me the same smack in the lungs. It still helped, but suddenly, the drama was gone.

In a few months, it seemed to become less effective, and I started relying more heavily on Albuterol, while still taking the Serevent every day.

Next time I go to the doctor, he asked why I'm not taking steroids. I told him I didn't like the taste. He said a new class of drugs had been released that didn't seem to have taste problems for most people.

It was a non-aerosol inhaler that had no puff. I wasn't even sure if I was getting any medicine when I took it, but it seemed to help. I still hit the Serevent, and didn't feel quite right if I didn't take some regularly. As I generally saw different doctors, they listened to my regimen and tried to work with it.

A call came from the HMO one day. The nurse said I could see various doctors, but I should choose a primary care physician (a decision I'd been avoiding for months). They said I wouldn't be able to see any more doctors unless I signed up for primary doctor.

I picked one almost randomly, but probably wisely. After dealing with my asthma rather haphazardly, some discipline was about to be instilled.

Chapter 18

The Authoritarian Doctor

My wife went to see the Authoritarian Doctor first. She didn't like him immediately and chose a different primary physician. Upon meeting the Authoritarian Doctor, I could see why.

The Authoritarian Doctor appeared overweight, grey-haired and rather brusque. He sometimes waved questions away with his hand. He didn't have the same smokers' hack, but he reminded me of my first pediatrician, who had seen me until about the time I entered my teens.

But he seemed to be what I needed at that moment, so I chose him as my primary physician at the HMO.

If I really couldn't stand him, other doctors could replace his primary physician status.

If I was to be his patient, the Authoritarian Doctor announced, I *would* take the inhaled steroids that *would* get my asthma under control. If I smoked cigarettes (I tried to say I didn't, but he ignored my interruption as if I said nothing), I *would* quit. He was tired of seeing smokers who only made their problems worse; if they weren't willing to quit, he wasn't willing to treat them.

Even in my late 20s, with only a hint of a rebellious streak left, all this speaking in the imperative rubbed me the wrong way.

But, when I had seen him before, he refilled the prescription for the Serevent (which was what I really wanted), though he said the inhaled steroids were more of a first line treatment, while the Serevent was more of a second line approach.

He didn't like the steroid inhaler I used – he recommended another called Asthamcort, and it seemed to start working (though I didn't think of cutting back the Serevent).

The Authoritarian Doctor said appointments should be scheduled every three months until the asthma came under control. That's what it was about, he said, "Control, not crisis management."

He became my doctor and insisted on a full check-up. It was the first one I'd had for years. He said I should keep an eye on my cholesterol. He recommended a diet that disagreed with me so violently that I think it helped to give me a hernia (which he diagnosed with efficiency).

But, after three or four visits, the asthma seemed to get better. The idea of control was becoming more clear. The Authoritarian Doctor insisted upon my attendance at a class about asthma management.

There I was told to use a peak flow meter – a device which measures lung power at the moment. And, they also said indigestion problems were sometimes related to asthma, but doctors weren't sure exactly how the two were connected.

This was fascinating to me, who burped like crazy after most meals, and sometimes felt a bit sick to the stomach without really knowing why.

The possibility of other drugs also intrigued me while scaring me a bit. The Authoritarian Doctor said I should be taking antihistamines every day, even when I didn't need them. He recommended Allegra, though I had been taking Claritin, which seemed to work fine for me. He gave me a pill called Singular that he said had a synergistic effect with the Allegra.

I wanted to impress the Authoritarian Doctor with my obedience, so I tried to follow the regimen. A puff of Albuterol was usually administered right before my appointments; I wanted clear-sounding lungs when the cold metal of his stethoscope hit my chest.

After preferring interesting and erratic spontaneity to planning, and free-spiritedness to cold order, I slowly got a feel for the value of methodical treatment, and for those who smugly insist they know best.

The Authoritarian Doctor may or may not have known best, but he certainly knew better than me. He had wisdom and insisted on respect for it.

I learned to do a better job of controlling my asthma, but also that some people have knowledge to share, but you might have to work with them on their terms to get that knowledge.

Chapter 19

Almost Cold Turkey

Visits to the Authoritarian Doctor continue every three months, as he had insisted.

Each time, I seemed closer to being stable, but not quite there. There was sort of a routine, but at some point, he seemed to show some disapproval when I asked for a refill of Serevent.

He still refilled the prescription, but maybe it was time to get off that particular drug. Strangely, I didn't really consider asking the doctor how to safely stop taking the medicine – to do so was to acknowledge that something had to be done. And, even if something had to be done, I figured I could handle it.

By that time, Serevent definitely didn't offer the blast it first had, and the inhaled steroid really did seem to be having an impact. For a day, I just stopped using Serevent. That didn't work out that well, as I felt the quality of my breathing diminish before dark.

I decided I should try to taper off about the time I was taking a trip to Florida, where my parents had moved. Sometimes it seemed that Florida air wasn't quite as conducive to asthma problems as the air in Illinois.

Right before the trip, I cut my Serevent dose from four puffs a day down to three; after a few days, I went down to two. With two days left before heading home, my dose was reduced to one puff a day.

The urge to take more was powerful, but I also knew intellectually that the steroid and plain Albuterol worked, so the schedule was maintained.

Two days before the end of the trip, I stopped using the medication. The next day my lungs felt bound with gunk I simply couldn't cough up.

Borrowing my Dad's rickety bike, I started pedaling slowly through the mild Florida air to get my breath moving. As I increased the pace and my inhalations increased, I could feel the passages inside my chest slowly opening.

It felt like the end but I didn't sleep well for a few nights. Within a week, however, everything returned more or less to normal.

At the next visit to the Authoritarian Doctor, he didn't even seem to notice the absence of a Serevent refill request.

Despite his indifference, I was glad to be off the drug. Over time, the medicine's effectiveness really did seem to disappear, leaving me with a desire for it, but no real positive impact.

Years later I wasn't surprised to read articles which described asthmatics dying of attacks while clutching inhalers with similar drugs in their hands, and concerns from some physicians about the safety of such drugs – though they are still available today and most users find them safe and effective.

While the Authoritarian Doctor was teaching me the importance of compliance, I still kept a skeptical eye out, and the Serevent incident showed me why that skepticism can be helpful.

It wasn't just about taking the drugs on schedule, it was about taking the appropriate drugs on schedule.

Section Five

Stern Teacher

Chapter 20

Blood

After my triumph of self-regulation in the case of the Serevent, I got a little cocky.

Though the Authoritarian Doctor told me to keep on taking the medicines throughout the year, I needed them less in the winter and summer, when pollen is relatively low.

After the winter freeze, I felt a little better, and stopped taking antihistamines. Then Singular pills were cut out of the routine. I didn't stop taking the steroids, but the schedule became lax.

The Albuterol remained near to me and was used on certain occasions.

A winter chest cold developed, eventually creating a whistling sound when I breathed. I woke up with coughing spasms.

Continuing to define it as a chest cold which really doesn't need medical attention, I was quite startled to see some streaks of red in the gunk which I cough up violently.

I get out the American Medical Association's Family Medical Guide and look up the possible causes. The only thing I find "blood in the sputum" associated with is tuberculosis.

Oh dear, I'm going to get sent to a sanitarium, hopefully before I've infected my family. And then I have a worse thought. What is the Authoritarian Doctor going to say when he hears:

a) I'm not keeping my symptoms under control; and

b) that I've unilaterally decided to cut back on medication.

I should have called the doctor's office right away, but I wait, thinking about what I'm going to tell the receptionist. She's actually not a problem as she tells me that the Authoritarian doctor is out of town for the week. I could wait, or I could see another doctor in the HMO that day.

It seemed best to go right away.

At the doctor's office, during the nurse's interview, she asked if I had been taking the list of medications I had been prescribed. Uh, not all of it I said, before elaborating.

When the doctor, who was younger and much less stern looking than the Authoritarian Doctor, arrived I told him about the blood, but refrained from offering my own diagnosis.

He looked at me with an even sterner look than that Authoritarian Doctor usually mustered and asked, "Why haven't you been taking your medicine?"

I told him about my seasonal reactions, and how I usually don't take it during the winter. He barely let me finish.

"You've got to take the medicine," he said.

He also gave me some antibiotics as he thought I might have had an infection. The only light moment was when I finally asked him if it was possible that I had TB.

His lips almost made a smile as he shook his head in the negative.

"*You've got to take the medicine,*" he repeated with emphasis.

I would keep up on the medicine. Despite my earlier pride at being in touch with my own body, it was time again to let the experts make the big decisions.

Chapter 21

Steps Forward – Steps Back

After my lungs recovered, I followed the doctor's advice. The steroid inhaler was puffed with almost religious regularity – twice in the morning, twice in the evening. My mouth was rinsed after each session to avoid thrush and other infections of the mouth.

Allegra and Singular were taken every day, and the Albuterol as needed. My life became structured around the words written on the sides of medicine bottles. And, at first, my need for the Albuterol went down.

For several months, everything worked more or less as it was supposed to, until the next fall, as the ragweed released its pollen. More Albuterol was needed, and even as the season ended, the need didn't go away.

The Authoritarian Doctor suggests increasing the dose of the steroid – up to 3 puffs twice per day. If it doesn't help, he says somewhat gravely, we might try oral steroids, something which didn't sound appealing.

Upping the dose of inhaled steroids helped, making the Albuterol less necessary.

I thought my allergies were getting worse. Even though antihistamines were taken every day, my nose felt congested all the time, and I experienced terrible indigestion regularly, after almost every meal.

My asthma was stable again, but I just didn't feel too good. When springtime came, the bad feeling intensified.

I read about massive fires in Mexico supposedly affecting people with respiratory problems hundreds of miles away. Could the pollution have made it all the way up toward the Great Lakes region? After the fires disappeared from the news (I'm not sure when they really went out) I seemed to feel a little better, for a while.

Some days were fine, and some just awful. Sometimes one bad day would turn into a week. If I tried to clean and stirred up dust, that made it even worse.

Frustrated and a little frightened, I knew that taking medicine for years (possibly a lifetime) was part of having a chronic illness. But the increasing need for more medicine to achieve the same effects was disheartening.

It felt as if the asthma and the medicine were the powerful forces, while my own body was just a battlefield, and all I could do was passively watch the battle.

When browsing bookstores, I began to look at titles that questioned conventional medicine. After buying some books about asthma which focused on alternative therapies, I only peeked at a couple pages (diet, exercise, breathing techniques, all, to my mind, inferior to the technological force of the pharmaceutical industry) before placing them on my "to be read at some time in the future" shelf.

A couple times it seemed the problem might have been Albuterol itself, so I tried to stop inhaling for a while. This usually lasts a little less than 24 hours, until I frantically gasped at the healing mist released from a simple press of the puffer.

Then I felt better again for a few hours, until the burning, acidic feeling in my stomach would wake me right about 3 a.m.

Now conscious of my strained breathing, another Albuterol puff is taken, which brings enough relief for sleep.

This becomes an ongoing pattern. Afraid to see the Authoritarian Doctor because I don't want to step up the medication, I soon find that's one worry I need not take seriously.

Chapter 22

The Disappearing Authoritarian Doctor's Amiable Replacement

The medicine was running out. The Authoritarian Doctor only renewed them for three months at a time, and it had been about three months. Feeling a little worse than the last visit, but not bad enough to change course, I was prepared to argue against increasing doses of medicine.

During the call for an appointment, the receptionist said the Authoritarian Doctor wasn't taking appointments. She couldn't tell me why or for how long, but she could set an appointment with another doctor, which is what she recommended.

A new doctor had joined the staff at about the same time. Young (at least to my mind, as she seemed to be about my age) and seemingly open-minded, she was the doctor I saw that day. After entering one of the rooms where the doctor sees patients, I asked the nurse what had happened to the Authoritarian Doctor.

She hesitated, and then said she wasn't quite sure, but there was a memo that said the disappearance should not be discussed with patients. That struck me as about as honest as she could be, but the thought was forgotten as the doctor entered cheerfully. The young new doctor was quite amiable, and I automatically liked her more than the Authoritarian Doctor.

The Amiable Doctor asked about my medication, what seemed to work, if anything didn't seem to work. I told her all about my regimen, and said I'd felt a little down. She recommended one more increase in dose of the inhaled steroid, which put me at the maximum dose.

She didn't treat it like a big deal, but she said it was worth a try. She made it sound reasonable to me.

Looking over my relationship with the Authoritarian Doctor, it wasn't always perfect. I had used one antihistamine, but he prescribed another. When I asked to change back, he essentially refused.

He was really stubborn about it; his resistance made me think he may have been getting kickbacks from a pharmaceutical company. However, he helped me with other health problems, and he did really show me that my asthma could be kept relatively well under control with consistent use of medication.

But I was deferring to his judgment for the most part. It seemed time to share control cooperatively.

Though I never learned the reason for his departure, it came at an appropriate time. It may have only been subconscious, but I started thinking about self-responsibility. I started to page through those asthma books collecting dust on my shelf– not to challenge doctor's advice, but to augment it.

I was free to take responsibility for myself, but I didn't really understand how important it was at the moment. Asthma was about to hand me another uncomfortable opportunity.

Chapter 23

Cornography

We had been sightseeing in Florida, where my parents now lived. With my wife and kids, we spent the day at a state park looking for alligators. We were hungry, so we stopped at a seafood restaurant we often visited.

My dad ordered Calamari, and corn muffins came out along with it as appetizers.

I enjoyed the taste as I always did, but a few minutes after the first bite, my lungs clenched closed. A couple puffs of Albuterol returned my equilibrium, but the corn-encrusted squid completely lost its appeal. Later as my stomach burned with indigestion (despite taking a couple of chalky antacid tablets) there was time to think.

Food and asthma were related. Previous attacks sometimes seemed vaguely related to eating, but I couldn't recall any reaction as obvious as this one.

There was a time when cornmeal pancakes were made on Saturday mornings as a weekend treat for my wife and kids. I had to use the Albuterol inhaler heavily many of those afternoons. One Saturday afternoon my wife drove me to the doctor's office glancing over at me nervously as I strained for breath in the passenger seat. The maximum dose of Albuterol had been taken and relief was still only partial. That day, the doctor in the office asked why I hadn't come sooner as the nurse prepared a special nebulizer.

The cornmeal pancake breakfasts were discontinued years before I thought I might have a broader problem with corn. It made sense that something in the mix was making me sick, but the main ingredient itself didn't really cross my mind.

Then, there were the movie theaters, where a puff was always needed about halfway through. Only a few months before, at Halloween time, I remember a corrosive sensation in my stomach and the sudden need for a puff of Albuterol after eating a small bag of Skittles, which were loaded with corn syrup.

I decided to stop eating corn to see what would happen. Obvious stuff, like whole kernel corn, corn bread, corn flakes and others products which clearly contained corn were avoided with general ease.

I felt a bit better, but soon learned corn was hidden a great deal of my daily food intake – from soda pop to frozen dinners.

From corn starch, modified corn starch, high fructose corn syrup, corn oil to more esoteric chemical concoctions, like maltodextrin and sucrose – it was all made from corn. Read the label of most candy, snacks and prepared food and you'll see the many variations on corn. The damn antacids, which I thought were helping me, were held together with corn starch.

I loved candy, snacks and prepared foods. I hated asthma and all the medicine, so I decided to make a concerted effort not to ingest corn.

It wasn't easy. I became one of those guys in the grocery store who clogs up the aisles while he tries to read a dense paragraph of ingredients on the Lean Cuisine label. During the first months my fellow shoppers might have heard me cursing under my breath as I learned of another thing I couldn't eat.

The food I could eat, however, was good: fresh vegetables and fruit, meat, grains and a few trusty candies, like Heath Bars. After reading that the regular intake of some fruits like apples may help to control asthma, I start living the adage of (at least) an apple a day.

I lost weight. Going out to eat became tricky – sometimes items with hidden corn were mistakenly ordered. Then, about halfway through the meal, the constriction was felt. Pushing aside whatever food seemed to be bothering me, if it wasn't already gone, a dose of Albuterol helped.

After a few months, the progress continued and I was excitedly telling people about avoiding soda pop as if it were poison. I didn't give up any medicine, but for the first time in years, on some days I felt the Albuterol wasn't necessary more than once a day.

But some days were still better than others.

Chapter 24

Beyond Cornography (Yeastography)

I didn't feel sick all the time anymore, but I didn't feel well all the time either. Sometimes after meals my stomach just didn't feel right, despite my certainty that corn was absent.

I read further in the books I had bought about asthma, and sought out more books with an alternative theme. Most didn't mention corn as a problem. One obscure book noted corn was a major problem for many people, but the book also said asthma and allergy problems could be cured by holding vials of one's own spit in one's hand for a designated period of time.

The alternative world had been entered, where things seem to make sense at first, but then get weirder the more you hear about them.

Some alternative books said chronic stomach problems could be due to yeast overgrowth of the digestive tract, or Candida. It wasn't recognized by mainstream medicine, but several alternative sources describe the malady, and whole books had been written about it.

Special diets without sugar were recommended, but I thought maybe yeast was a problem. The books said people who took inhaled steroids sometimes had the yeast overgrowth, theoretically because the steroids killed other bacteria typically present in the digestive system.

It kind of made sense, and the Amiable Doctor wasn't totally dismissive when I asked her about it, but she wasn't too impressed either. She said if not eating corn helped, then I shouldn't eat corn. Same with yeast, but she said she wasn't aware of other people with major yeast problems.

So I also tried to give up yeast. Not only is it in most breads and crackers, it is also processed for use in many prepared foods. And good convenient starch like pizza and doughnuts, forget it.

But further refining my diet didn't bother me all that much. When I started trying to avoid yeast, my stomach started feeling better, and I stopped daily use of the Albuterol (though it was sprayed every few days). I continued on my other medications and tried to read deeper on the subject.

Some things I've read before also come back. In John Updike's memoir *Self-Consciousness*, he describes his overuse of an inhaler, and finally figuring out that he's allergic to the family cat. Instead of forcing the cat out, it's Updike who leaves, telling readers his family would miss the cat more.

That's pushing it more than a bit, but maybe it was best to make as clean a break as possible.

Sometimes, I'd make a food mistake. At a party at a friend's house I was offered a mixed drink. After my second one, I noticed an itchy sensation in my throat and a tightening in my chest. I went back to the kitchen to see how the drinks are made - primarily with soda pop and alcohol.

Incidents like that strengthened my resolve to stay away from the things that bothered me. At first I sort of craved them, but then, like a reformed cigarette smoker, that shiny, lightly brown caramel covered popcorn stopped looking appealing. I finally started to understand that the very short taste pleasure of the junk food was not worth the discomfort that followed it.

It wasn't always pleasurable, but the experience taught that life sometimes forces big choices and careful attention, which can have significant payoffs.

Chapter 25

Back To Basics

Though some of the asthma books had been paged through lightly, I hadn't even finished one completely. The diet success, which may have happened years earlier if I simply took the time to do a little research, pushed me to learn more.

There were several books to choose from but something about the copy of *The Asthma Self-Help Book*, which I still hadn't read, appealed to me. It was the first asthma book I bought many years ago. First published in the early 1990s, the book seemed a little dated, but I soon realized it featured easy to understand explanations of asthma, along with possible causes and its treatments.

I knew some of what the book discussed, while other information was new to me. Author Gerri Harrington, who discusses her own asthma in the book, makes the candid admission that asthma is in many ways a mystery right up at the front.

On the first page of the first chapter, the author notes that asthma is increasing and nobody knows why. The book also says the underlying causes of asthma are unknown.

Harrington expresses belief in the individual's ability to control their asthma, and not only through drugs. On the second page, she talks about diet and exercise; there are chapters about breathing techniques, stress-related asthma and stress reduction to prevent attacks.

In the years between the book's publication and the time I finally started to read, some of the information seemed stale, but the tone and breadth seemed totally fresh to me.

Unlike some of the alternative-based books I read, it seemed quite balanced between the need for traditional medicine as well as more holistic approaches.

Occasionally, a Zen-like pronouncement popped out of the text.

"Self-care … means being neither resistant to taking drugs nor totally dependent on them, but rather fitting them into an overall plan of good health."

Then it clearly explained the virtues and risks of drugs I'd been taking for years. One of the chapters is titled, "The Pros and Cons of Corticosteroids."

As soon as I finished the book, I thought quite a bit about it, but then kept on doing what I had been doing. Avoiding corn and yeast seemed to be working (though they weren't really recognized as causes in the book) and I was feeling not too bad at the time.

In a sense, I had too many possibilities after reading the book. I knew that there would be other books to read, and made a mental list of things to try: Yoga, breathing exercises, meditation, maybe even hypnosis.

Some of them I would go on to try in time; some of them I wouldn't. But Harrington's book is still my favorite that deals strictly with asthma. She has knowledge to share, and she does it in a friendly, conversational way. She makes you think that she's not all that much different than you, and the improvements that she experienced are possible for the reader as well.

Unlike going to the doctor all these years, it was more than just strict instructions. It was an effort to be thoughtful about the instructions, and to think about the implications of the instructions.

The doctors, I think, all tried to help, but they didn't really talk in terms beyond medication.

The book reminded me also that knowledge doesn't fall into your lap all the time – sometimes it must be pursued with no help, and if you don't try, you won't find it.

Chapter 26

Need to Learn More

The information in the Asthma Self-Help Book was considered, but not acted upon. In fact, coasting with the improvements from diet changes, I thought maybe I'd learned enough.

But then, my 3-year-old son, already diagnosed with allergies, started showing signs of a breathing problem. He probably displayed symptoms for at least a few days before I noticed.

The usually energetic boy was moving rather sluggishly; he wasn't talking much.

There wasn't really a wheeze to hear, but I finally got it one day when I saw him sitting on a chair on a nice late summer day wearing a strange look that combined exhaustion and frustration.

The poor kid can't breathe, I finally realized, more than a little embarrassed that I had started to consider myself something of an expert on asthma.

We took him to the doctor, who made the asthma diagnosis and prescribed a nebulizer machine, which turned medicine into mist to be inhaled through a tiny face mask.

My son didn't like wearing the mask, but he enjoyed watching a demonstration video that came with the contraption, so we established a ritual of playing the video at high volume to overcome the hiss of the machine while inhaling the mist.

The treatment succeeded and the symptoms soon came under control.

The episode showed me how much I didn't know about asthma, but it also brought back the flood of memories about having asthma as a child, and concerns about what my son might have to go through.

I would need to learn more about asthma. And, I thought somewhat selfishly as a writer, maybe it'll lead to an article, perhaps even a book.

My local public library had dozens of books related to the subject. I checked out about half of them.

My reading was rather haphazard, but a schism seemed apparent. The conventional medical perspective relied primarily on pharmaceuticals; while the alternative perspective took many approaches, but was not as enthusiastic about, and was indeed sometimes hostile to, pharmaceuticals.

My expectation that the medical books would give precise causes for asthma, and the alternative books suggest more nebulous and esoteric causes was wrong. I had it in reverse.

The science books were pretty clear: The symptoms of asthma are diagnosable, but the causes "remain uncertain."

Some of the alternative books, however, seemed pretty sure why asthma troubles so many (diet, general lifestyle, pollution, etc.).

Diving into the subject, I expected to get a coherent view of asthma. But the diverse viewpoints, caused my own view to become more fragmented.

Chapter 27

I Briefly Think I'm Having a Heart Attack

My strategy to deal with asthma at that point might have been called the regular therapy plus push back method.

I took my medicine regularly, tried to avoid what bothered me.

When the problems went beyond the capabilities of the regular regimen, I used one of the other tools I had acquired over the years: either the Albuterol rescue inhaler, or exercise, or a glass of water - whatever seemed appropriate at the moment.

If the asthma pushed me, I pushed back.

One evening in the late summer, just when pollen was starting to get bad, I went for a run, despite feeling a little tight-chested. Just as it had back in junior high, that could still force more air through my lungs. But as the run started, the tightening sensation got worse, not better.

Puffing at the Albuterol slowed my pace, making it an effort to not lapse into a total walk.

As it usually did, the Albuterol helped my lungs to expand so the air started to flow. Feeling a little better, I ran faster.

With headphones on, pushing to run with the beat, I felt a sudden sharp pain spread from the top of my shoulder to the upper right quarter of my body. Trying to run it off for another minute, the pain did not fade away.

The rest of my body felt fine; my breathing was great, my legs seemed ready for a workout, but the pain surging around my shoulder just didn't stop. Is this what happens when you have a heart attack, I thought just before stopping. In my mid-30s it was unlikely (though not impossible) for me to be having a heart attack. The discomfort receded a bit during the slow walk back home, but by later that night, the feeling still lingered.

By the time of a doctor's appointment a few days later, the pain was gone, but not the scary sensation that something bad had happened.

The Amiable Doctor listened to the story seriously, checked all my vital statistics, but ultimately said it wasn't much to worry about.

She said when the lungs are straining particularly hard, and possibly expanding beyond their normal range, they can push out against shoulder and back muscles, as well as bones, causing pain. She checked to make sure I was keeping up with my regular medication (I was). She noted that if other problems arose, the steroid dose could be raised, and that when allergy season was over, reduce the dose back down.

It seemed to me that the asthma has pushed me harder than I could push back. Not thrilled about taking more medication. I decided to research more in depth to find other weapons for my grudge match.

Chapter 28

Stretching In Other Directions

Being a writer, I envisioned researching some of the alternatives, and then writing about them. Not quite sure where to start, yoga seemed like one therapy endorsed by both alternative advocates and mainstream medicine. I decided to give it a go.

The local park district offered a weekly class for beginners. Though one of the only new people in the class, the teacher and other students were very welcoming and helpful, explaining and demonstrating the positions my body was supposed to reach.

The lessons always started with stretching and breathing exercises, followed by the actual attempts at the poses. My breathing seemed a little clearer after the first few classes, but the difference was not dramatic.

The class, however, did lead to one invaluable insight. The yoga instructor always stressed breathing from the diaphragm, the muscle right below the rib cage, but it was difficult for me to tell if that's what I was really doing.

During one of the classes, we learned the cobra pose, in which you lie on your front, torso lifted, head looking up. When we got in the position, the instructor said it was a good opportunity to feel diaphragmatic breathing, because the diaphragm was the part of the body pressed to the floor right before the upper half of my body lifted off the floor.

And I felt it, realizing up until that point, I probably wasn't pushing my lungs wholly with my diaphragm – instead using shoulder, chest, stomach muscles. But just then, it felt good to press the diaphragm and feel my lungs empty more completely, leaving me to believe they would fill more completely too.

The teacher of the class frequently offered interesting observations about yoga, breathing and general health.

I only took the yoga class for one session. Later, a DVD and book were used to refresh my memory. When I'm feeling tight, I still sometimes pull them out, or at least try one of the poses I remember. I could probably have learned more, but I was ready to see what other therapies could teach.

Chapter 29

Writing For Breath

While traveling on vacation with my family one summer, disappointed by the reading fare available at the moment, my wife suggested writing a better book. So instead of just writing an article about yoga, I'd try to tell more of a comprehensive story on asthma, and see if it helped along the way.

This meant more research was necessary.

Mainstream asthma texts continued to offer similar information about standard drug treatments. Some alternative books were very similar, offering the same sort of suggestions as my old favorite *The Asthma Self-Help Book*.

And then there were different kinds of books, some that looked at asthma in a whole new light. Some were memoirs by asthma sufferers; some were historical pieces; others were new-age tracts about getting in tune with the universe; and some offered speculation about the connection between spirituality and health.

Initially, overwhelmed by the scope of all that had been written, my reading and note-taking was haphazard. But, recognizing ideas and anecdotes about asthma from others that resonated with my own experience, the project started to get under control. Strangely and simultaneously the asthma also seemed a little bit more under control. Both my mind and body were looking at the issue intensely; perhaps my asthma itself was a little intimidated by the scrutiny.

My first round of reading suggested different ways to look at asthma. Some books insisted the problem wasn't so much what asthmatics breathed (pollen, dust and pollutants) as how they breathed (either too deeply or too shallowly). Some saw it strictly as a problem of diet. Others looked at asthma as an almost inescapable reality, an invisible but physically uncomfortable manifestation of modern life's crushing quality.

Sometimes these books caused personal revelations, sometimes they evoked my derision. But I had to acknowledge, whatever asthma was to me, it might have been something completely different to someone else. Asthma is an objective fact, but its ultimate nature remains a question of perception.

I tried to reconcile that information in my mind and in my daily life. General health seemed to have a connection with asthma. More physical, but non-medical, changes could be made.

Some of the changes were completely fundamental. For example, after obsessing about what I ate, it was finally time to consider how I ate.

Chapter 30

Relearning the Basics

A food wolfer who ate too fast since childhood, my bad habits intensified in college while working as kitchen staff at a sorority. We set the tables, brought out the food, cleared the tables and then washed the dishes for the sorority sisters. In between serving and clearing, for roughly ten minutes, we ate too.

At that age, I ate a lot, consuming tons of food while barely chewing it. It's a habit that stayed with me past college, eating too much too fast.

After seeing first hand the connection between eating and breathing, I paid more attention to sources that focused on the subject.

One book suggested eating healthier foods and chewing them completely. Well, duh… Then an image of my grandfather, who lived into his nineties, popped into my head. He was saying each mouthful of food should be chewed 40 times. I don't recall him ever complaining of indigestion or breathing problems either.

But, it was hard to accomplish in actual practice. Used to rushing through meals, unless I concentrated, slowing down was hard. Until my chewing became more disciplined, one book suggested eating enzyme supplements with meals to aid digestion.

Over time, I stopped feeling less full and congested after eating. The slowing down and the enzymes really seemed to help. I asked the Amiable Doctor about the enzymes, and she said she was skeptical about why they would help, since they should just break down in stomach acid like everything else. But, if the enzymes helped, she said, there was no reason to stop taking them.

While working on eating slower and more consciously, I tried to do the same with my breathing.

Many books suggested focusing on breathing diaphragmatically, as I had learned in yoga class. It was about using that big muscle underneath the lungs to press the air passages completely clear of air, and then moving the muscle in reverse to open the passages fully for easy air movement.

I really pressed up my diaphragm and squeezed every square centimeter of air out of my lungs. Those organs probably had not felt as refreshed since junior high cross country.

Diaphragm practice took place very consciously for a few minutes a day, in an effort to train myself to get used to it.

Again, it didn't come naturally, but really thinking about it (and being forced to think about it as I read) seemed to make a difference.

Even now, years later, the bad habits start to creep back, almost imperceptibly. When the problems start to show themselves, then I have to live more consciously, and think about the things that are usually done unconsciously.

Chapter 31

Changing Your Life Up to a Point

Asthma and allergies are often aggravated by dust, and my experiences indicated a clear reaction to dust. But there was a problem. I'm just not much of a housekeeper.

I've improved since my teen years. I tried not to let the mess in my office or elsewhere in the house get to unsafe proportions, but a better effort could be made.

Many books about asthma recommended sanitizing one's living space, so I started. But more cleaning seemed to expose and stir up more dust.

Changing my life seemed like a great idea, but the confusing part was where to draw the line.

Being a writer, I keep lots of old books and papers around the house. They all collect dust; some even generate their own dust. Fun for me often involves going to old dusty book and record shops, and sometimes bringing the dusty books and records into my home.

Cutting down on the paper or getting rid of some old books was considered, but I decided not to do it unless asthma indicated that such a change must take place. Of course, that could be my own rationalization; perhaps I've missed the message all along, and the asthma has simply been telling me in the imperative: "Clean up your life!"

Maybe not, though.

The general messiness may be unjustifiable, but the old papers, books and records seem to offer some recreation and joy for me, and are therefore worth the risk. If I had a stronger reaction, say on par with corn or ragweed, the decision to not avoid these obsessions might have been harder.

In some ways, asthma helped me to sort through what was important to me and what wasn't. And, it seemed like it was more important to get more old books about asthma than to avoid the dust from those old books.

Asthma may have been telling me to write a story – I was about to learn that might have been the objective all along. And if it's not the objective, it's a perfectly good reaction.

Section Six

Contemplation Object

Chapter 32

Authoring One's Own Story

Read about any specific subject, and certain influential titles are cited repeatedly. Reading about the psychology of chronic illness, I saw several references to a book called *The Wounded Storyteller* by Arthur W. Frank.

After finding a copy of the book, I read it hungrily. Right in the preface, the author challenges typical conventions about what it means to be sick:

"I hope to shift the dominant cultural conception of illness away from passivity – the ill person as the 'victim of' disease and then recipient of care – toward activity. The ill person who turns illness into story transforms fate into experience…"

Whoa…

The author, who wrote the book after surviving a heart attack and testicular cancer, offers academic analysis of the problems of being a patient, and something of a solution.

He says stories of illness are largely dictated by the medical establishment, when there are other narrative approaches that can be followed. Indeed, sometimes whole new literatures should be formed by sick people themselves as they face illness.

The book doesn't suggest that illness will be automatically overcome if the ill start telling their stories, or that not facing a situation seriously will make it go away. Sometimes it's just a way to cope and seek meaning. But, sometimes, telling a story honestly can have a therapeutic effect. And moving from simply relaying the story to envisioning where the story goes represents another step in the process.

Telling the stories isn't just beneficial for the people who are sick, and others who face similar situations. Frank quotes many people facing illness, including a chronic pain patient who suggests that while doctors often try to educate the sick, her own experience offers an educational opportunity for doctors and others who don't face sickness: don't waste the precious good moments of your life.

Frank raises provocative questions not only about accepting and surviving illness, but in learning from it, and seeing it as not only a curse, but an opportunity.

"The problem truly is to listen to one's own story, just as the problem truly is to listen to others' stories," he writes, capturing some of the mystery of illness and redemption.

Though asthma isn't specifically mentioned in the book, Frank offers a whole new perspective about what happened. Writing this book and simultaneously learning about different ways to look at the world and myself wouldn't have happened without asthma (though I'm certainly glad it wasn't another nastier chronic disease that helped to teach me).

In my pile of books, I'd seen a few other stories about people who may have viewed their asthma as a curse, but then went on to achieve success in part because of the limitations asthma placed on them. And I was starting to see some of those in my own life.

I kept looking for stories that would help me to understand my own story. But it may have been my own story that finally helped me to understand others.

Chapter 33

Overcoming Asthma, or Using it to Overcome?

Search the internet for "famous people with asthma," and you'll get a long list. Often, the lists are presented by asthma foundations as proof that asthma can't stop people from success.

And it's true, asthma did not stop them from success. But after reading *The Wounded Storyteller,* and some of the other books, I began to wonder if asthma helped some find success.

Martin Scorsese, the great film director suffered from asthma as a child. Instead of playing outside with friends, he spent many afternoons in movie houses, getting the education he needed to be a master.

Likewise, Jim Davis, creator of *Garfield* (the comic strip cat/marketing powerhouse), spent time drawing as a young person because it seemed like a quiet activity that would not stir up his asthma.

Sometimes people who don't become successful because of asthma, some who even hate it, acknowledge the opportunities it presents. In his book *Catching My Breath,* Tim Brookes constantly curses asthma and the treatments it entails. Yet, one morning, woken early by an asthma attack and forced to search for his inhaler in a car outside, he sees a beautiful sunrise, and even he acknowledges that only the asthma allowed him to witness the scene.

For years a copy of *Swann's Way* by Marcel Proust sat on my bookshelf, unread with other classics that I'd meant to finish. But in my asthma reading, I learned Proust suffered from asthma and reportedly wrote about it in his mammoth masterwork.

I decided he might have something to say to inform my own project. Reading the first few pages of *Swann's Way,* I understood it was only the first of seven volumes, and at about 600 pages, slim compared to other installments in the series.

It was a commitment, perhaps a year or even two of my reading time. But right away, I saw a few fascinating insights about asthma and health in general, and decided to finish all the volumes even if there were only a few more insights.

Proust himself only started to write seriously after he became bedridden by illness including asthma. Would he have ever started writing without the sickness? Would his insights be as sharp without his senses made acute by illness?

Would I have ever taken the opportunity to read about them if I didn't have asthma?

It was sort of like the Zen saying, in which a student says that before he was enlightened, a mountain was a mountain. But when he found enlightenment he realized the mountain was not a mountain. After enlightenment, the student again perceived the mountain as a mountain.

I still can't claim any sort of enlightenment, but at that point, I wasn't sure if asthma was still merely a medical condition, or if it was something else altogether.

Chapter 34

Into the Weird Stuff

Working through the collection of asthma-related books, I kept those that seemed a little more off kilter toward one end of the shelf. Some sort of made me laugh a little at first glance, but it was time to approach the books with an open mind.

Some information (particularly on the Internet) was difficult to take at all seriously. My favorite was the city in India where thousands of people made a pilgrimage each year to eat tiny live fish that were supposed to chase the asthma away.

Several books put a time value on the alleged therapy, with titles like *The 30 Day Asthma Cure*. I hadn't been too impressed skimming through the books, though I never really tried to follow the suggested regimens seriously.

One of those books, *Asthma-Free in 21 Days* by Kathryn Shafer and Fran Greenfield, carries the subtitle "The Breakthrough Mindbody Healing Program."

The book describes the connection between the mind and the body, but not in way that suggests asthma is all in the sufferer's mind. The book offers a number of strange (to my mind at the time) exercises designed to help the problem without more medicine.

After reading a while, I tried one of the exercises, which the authors called "The Golden Inhaler." While having difficulty breathing, an asthma sufferer is supposed to envision a golden inhaler and take puffs from that, before, or instead of taking puffs from the regular inhaler.

I thought about it for a while, and the next time my chest felt tight, instead of reaching for the real inhaler, I tried to think about the golden inhaler. It seemed a little silly, but I thought about how it would be effective, and then pretended to take a puff off the imaginary inhaler, following the same breathing motions I would have with a real inhaler.

After a moment, my lungs seemed to loosen, just a little. I tried it again, and they seemed to loosen a little more. It wasn't a miracle, and it wasn't like the discomfort just went away (a few minutes later the real inhaler brought more complete relief), but the exercise really did seem to help a tiny bit.

More exercises from the book were attempted, all with less success. Some of the other books at the weird end of the shelf offered other strange insights that were sometimes true. I learned that smoked therapy involving marijuana and other herbs was common before the advent of modern medications. I bought a 60-year-old box of "Asthmador Cigarettes" on Ebay after I read a letter from a woman printed in the local newspaper stating that it was very effective relief. I didn't try them, but I loved the unconventional approach.

Think about that – smoking as a therapy for lung disease. Of course, it's not all that unusual as medical cannabis makes headway into the culture. Indeed, as a clinically proven bronchiodilator, it's not that surprising that some people still smoke cannabis for asthma. I've had mixed experiences with that particular therapy, though many medical marijuana users prefer to use a device called a vaporizer that creates a steam-like mist without actual smoke.

Of course, this is not information that most doctors or even the strongest asthma advocates will share. It makes one wonder what other valuable information has been ignored or forgotten over the years.

Some people theorize that engines could run on water if oil companies hadn't suppressed designs in order to save their market. What if there was an asthma cure that didn't involve anything outside of the body?

Chapter 35

Discovering Breath Control

Butyeko. It's a word that came up in books and on the web. It seemed to be a natural treatment for asthma, but the descriptions never offered specific details. Breath control and control pauses were phrases associated with the technique, but the technique itself was elusive.

A book that purported to be about the method was at my local library, but I read the whole thing to find that it didn't really describe the actions involved, the book merely promoted the concept. To learn the specifics, one needed to approach a teacher, according to the book.

Searching around a bit more on the Internet, I found someone who said they would give away the secret. The control pauses involved exhaling, holding one's breath as long as one could, before inhaling again. Advocates of the method believe the longer you can hold the breath (the control pause), the better your lungs are functioning.

When I first tried it, holding my breath for only a few seconds felt really uncomfortable. But after relaxing, doing some yoga, and trying more, I could hold for about 20 seconds. And strangely, after the breath was released, I felt like my airways had expanded just a little bit. If I did a series of three or four exercises during an uncomfortable moment, and waited a minute or two, the effect wasn't much different from using an inhaler to open my lungs and let the gunk clear out – it just took minutes instead of seconds.

I started doing the exercises every day, eventually holding the pause past 40 seconds. There is much more to the Butyeko method, with great attention paid to nose-breathing as opposed to mouth-breathing, and I'm not even sure my technique is anywhere near correct, but it does seem to help me. Some other techniques that go by different names might also work. In Harrington's *Asthma Self-Care Book*, she describes "pursed lip breathing," which seems similar in process and effect.

According to its adherents (who are numerous in Australia and the UK) the Butyeko method works and they have at least a couple studies to indicate that it does. The late Dr. Butyeko, after who the technique was named, suggested that people get too much air in their lungs, and it is as bad for the body as overeating.

I don't know if this is true, and some medical professionals scoff at the notion behind the technique, but I don't scoff now.

After practicing the exercises for several months and watching my reliance on the Albuterol reduced to rare episodes, I decided to use the breathing technique instead of the inhaler next attack.

In the two years since starting, I haven't really looked back, except for one fall night when the inhaler came to my rescue when the breath control just wasn't enough. Indeed, I still carry the device in my pocket constantly. And, while weaning myself off the Albuterol, I continued to take inhaled steroids (while exercising, watching my diet and focusing on my breathing). But after a while, especially when there wasn't much pollen in the air, the asthma problem was beginning to be forgettable. I decided it might be time to reduce my reliance on other medications as well.

Chapter 36

It's All in Your Head

The title to this chapter may sound politically incorrect. Once a common explanation for asthma, the phrase now incites fury among some sufferers.

In *Catching My Breath* by Tim Brookes, the author explores his own asthma, but also the transition from the mid-twentieth century medical view that asthma was purely psychological. The notion was pushed to the extent that some physicians during the 1950s recommended "parentectomies" for young patients. The central therapy consisted of physically removing children from their parents' presence for recuperation. Without that psychic burden, the child was expected to improve. I suspect that would have made my asthma worse, but some children involved seemed to breathe better.

Regardless of what others experience, every asthma sufferer knows what they feel is real, and that they would not choose to make themselves sick. And indeed, today's conventional medical wisdom suggests that asthma is primarily a physical process

With some of the success I'd had with unconventional methods, however, I had to wonder.

A friend recommended a book called *The Book of the It* by Georg Groddeck. A contemporary of Sigmund Freud (who admired, and, some might say, borrowed ideas from Goddeck's work), Groddeck was a writer and physician interested in the connection between subconscious mind and body.

Written as a series of letters to a patient, a fictional doctor narrates *The Book of the It* with ideas, insights and anecdotes about illness and its origins.

The narrator, whose analysis of the medical and personal lives of his patients may seem callous to modern sensibilities, goes so far as to say, "Disease is a vital expression of the human organism."

When an individual can't obtain a desire consciously, something (the narrator uses the label "The It") within the individual uses the body on a subconscious level to help make it happen. Sometimes this causes great discomfort for the individual, but sometimes the body expresses what the mind cannot.

Does this mean that some people are making themselves sick? The narrator seemed to think so, but he didn't have today's scientific research which explains many physical processes. But could sickness still be trying to say something to individuals? Maybe.

I know intellectually that my own asthma attacks are usually precipitated by external irritants that enter my body through the respiratory system or the digestive system. And, now, in the fall when the heavy pollen brings sneezing and irritated eyes, I can't imagine what my body is telling me, other than "take some antihistamines, moron."

However, the times when my asthma was at it worst, it usually attacked at night, waking me from sleep. It may have been a completely physical reaction. But sometimes I wonder if something inside of me was using my body to say, "Stop slumbering through life - wake up and pay attention."

Chapter 37

Moving On

The next appointment with Amiable Doctor, I told her about the breathing exercises and my near total reduction in use of Albuterol. She said that was great, and to continue what was working for me.

I asked whether the steroids and other medications were necessary anymore. She said maybe not, but the medication should not be stopped suddenly. The doses should be tapered off slowly so as not to shock my body. She gave me a schedule to cut back over several days.

The tapering off involved slowly reducing the number of puffs taken per day – six (three in the morning and three in the evening) by that time. First, one in the morning was eliminated. When this didn't seem to cause any problem after a few days, I went down to four puffs per day.

Reducing the dosage down to two puffs in the evening, my body (or my mind) may have started to miss the steroids. I woke up a little wheezy one night after the reduced dose, but the breathing exercises took care of the problem.

The tapering down of the medicine continued without any significant problems. My breathing was doing fine, but I seemed to lose energy as the dose decreased. The energy returned later, but as the steroid dose diminished, my body seemed to wonder where they were.

Soon, it seemed to be just a memory. I was happy to be done with it, and thought about how a few years earlier it wouldn't have been possible.

The worst of asthma season had passed for the year, but I kept up with the breathing exercises, and kept track of my experience while continuing to read and write about asthma.

The strategy worked until early spring, when the fresh pollen floated back into the air. I'd stopped taking most medicines, even antihistamines, but after a few days of sneezing and having an itchy nose, I decided to go back to the antihistamines. That felt pretty good.

As the season progressed, a little wheezieness started occasionally, but disappeared with some control pause exercises. I had read that getting asthma under control might help to get other allergies under control too, but decided to be happy about the asthma progress while continuing to use medicine for hay fever.

In the asthma seasons since then, the breathing exercises have worked reasonably well. Heavy pollen is noticeable, but avoiding foods that bother me, working on breathing exercises and other types of exercise, and relying a little on antihistamines seems to be working fine.

I know where my emergency inhaler is, and it will be used if the situation arises.

The steroids could be taken again if it seems necessary in the future. I hope not to, and I'll take steps to prevent it, but seeing the power of that medicine, I have to respect its authority while trying to navigate its various implications.

My son still uses conventional medication, but I try to do some breathing exercises with him, particularly in season. He sometimes has a hard time seeing the value of the exercises. Of course, he sometimes he has a hard time seeing the value of the conventional medication, too.

Except for Proust, most of what I wanted to read about asthma was finished. Strangely, after the book research was done, a few days later I felt an asthma attack come almost out of nowhere.

It was like the asthma liked all the attention and wanted some of it back, even if it didn't speak with as much force as before. And when I gave it just a bit of attention in the form of breathing exercises, it seemed content to fade again into the background.

Chapter 38

Breathing for Health

Breathing exercises and other techniques have helped to control my asthma. Some scholars and researchers believe that controlling one's breath can help to control a wide variety of bodily functions and health problems.

When I first saw these ideas, they seemed quite far out, but after successfully accessing the golden inhaler and controlling my attacks through the control pause method, they didn't seem too weird any more.

One book, called *Ways to Better Breathing* by Carola Speads suggests that the quality of one's breathing impacts the quality of one's life. The author says the quality of people's breathing can be diminished by stress. Sometimes poor breathing patterns caused by stress don't go away after the stress subsides, according to Speads.

Throughout the book there are breathing "experiments" offered to help readers assess the state of their breathing, as well as for specific effects. Some exercises are designed to relax the breathing (and in turn the whole body), while others are designed to stimulate.

As I read, and tried some of the exercises, some seemed to do what they were supposed to do.

And just as proper breathing can help to improve life, another writer suggests improper breathing can lead to serious problems.

In his book *The Tao of Natural Breathing*, author Dennis Lewis tells how his own acute physical pain that couldn't be explained any other way was relieved when he started to practice Taoist breathing exercises.

Right near the beginning of the book, Lewis describes the importance of diaphragmatic breathing, and describes the diaphragm as "the spiritual muscle." The author claims that the diaphragm is impacted by emotional stress, which in turn affects the body. When that happens, people start breathing the "wrong" way, by raising the clavicles or shoulder blades.

It's the same diagnosis as yoga, but from a different perspective. And, I remember being a "shoulder-breather" before consciously trying to use my diaphragm.

Reading on, it was hard not to notice that following all the advice in the book would require serious time and study, which might have been worthwhile. But at that point, I thought of it more as a reinforcement of other lessons.

Are lots of breathing problems simply that, breathing problems? Certainly external factors can influence our breathing, but there seems to be something said for looking into ourselves as well.

But our breathing also connects us to the rest of the world and all its living inhabitants. The air that you breathe now has been breathed by your family, neighbors, and millions of people through history. It is the time when you actually draw the universe (at least a tiny portion of it) inside of yourself.

Indeed most people easily inhale the universe without wheezing – those who wheeze might just need a little therapy of one type or another. As described throughout these pages, many therapies exist, but attempting to find personal balance seems like a good way to welcome the universe into yourself.

Section Seven

Lingering Reminder

Chapter 39

Breathing Through the Asthma Epidemic

While I think asthmatics might want to look at the way they breathe, as well as the way they live in general, certainly those are not the only problems.

The development of asthma follows the development of industrial society in the world.

Here in the West, where reports of asthma increased dramatically through much of the late 20th century, before retreating slightly, asthma is as mysterious as ever.

However, it does seem to get concentrated in certain types of places – places without much economic hope, places with chronic environmental problems, places where people face hunger.

It's hard to determine exactly what asthma might be bringing to those already impoverished places, other than added awareness of the misery. But asthma also shows up in clean, wealthy places where food is abundant.

Is it just random?

For me, asthma has been an insightful if frequently cruel teacher. I don't know what it means to other people, or if it functions as a teacher for them, but it might be telling us that we have too much stress, too much pollution in the air, to much bad food in our diet.

Almost ironically, those of us who have asthma can't just think of ourselves as victims of a broken world. Indeed, as I write this, asthma sufferers around the United States are still using inhalers propelled by ozone-depleting CFCs, which certainly isn't helping environmental problems, though such inhalers are supposed to be completely phased out of use by 2008.

For some people with severe asthma, the experience is likely so horrible that it's hard to find anything positive in it.

I used to see it that way, but now after years of trying to look at the subject in another way, I can see asthma has helped me to start living the way I should be living. It forced me to pay attention to things I had ignored. It got me to read Proust.

And, finally, I finished Proust. The books were great, loaded with many insights about life and the world, but not as much about asthma as I might have hoped.

Then, right near the end, a small section of the last book hit me like a hammer. Proust doesn't use asthma specifically as his example, but this is what he notes of his own experience, and what finally forced him to write:

> "Ill health, which by compelling me, like a severe director of conscience, to die to the world, had rendered me good service (for "except a corn of wheat fall into the ground and die, it abideth alone: but if it die, it bringeth forth much fruit"), and which, after idleness had preserved me from the dangers of facility, was perhaps going to protect me from idleness..."

I first learned of Proust's asthma through a memoir written by Louise DeSalvo about her own asthma experience. While DeSalvo sometimes expresses rage at the suffering caused by asthma, she also clearly states that asthma has had value in her life.

If she didn't have asthma, DeSalvo wrote, she would "go out and find a way to *get* asthma," since she considered its attack and retreat as a very educational experience.

To me, asthma has been a teacher – one I wouldn't have paid mind to willingly – since as a teacher it's very demanding and sometimes quite nasty. And perhaps it's not finished with its lessons yet. But for now, it is a quiet and lingering reminder, both in my memory and my body, of the importance of attempting to be self-aware.

Chapter 40

What Really Happened?

I remember walking home after attempting to jog several years ago, when even the Albuterol couldn't help keep me going.

I thought to myself, is this what life's always going to be like? Will there be some miracle cure that comes along to restore me to normality?

But then, as I began to see some positive changes, it helped lead me to other positive changes. There was a sort of snowball effect.

Or at least that's the way I like to think about it in the context of this book.

Of course, some days, I'm not totally sure if my perceptions and actions helped to take control of the situation, or if it resolved itself through other means while I concentrated deeply on breathing.

In the midst of my improving asthma condition, I also moved from one house to another, and gave up some work I found stressful. Through the experience, my kids got older, and I believe the anxiety I first experienced as a new parent may have subsided continually, also reducing stress levels in my life. I started getting more involved with my church.

Therefore, I must acknowledge other explanations for my experience beyond what I've examined in these pages. Or perhaps, only one of the many regimens I tried was really successful.

However, I swear that what I think happened really did happen. That is to say, as I became more aware of asthma as just another voice in the body as opposed to an oppressive curse which could only be exorcised through the ceremony of medical science, I learned things that helped improve my overall life, not just my asthma.

Maybe your asthma (or something else in your life) is telling you that you should be utilizing all the latest and greatest technological innovations from the halls of medicine – I'm certain sometimes it does. And, I will acknowledge that medical science has some advantages over trying to think about the experience in a different way.

There is a paradox, not unlike the running dreams that used to wake me up as a child, in which the energy in my legs seemed limitless while my poor lungs grew tighter and tighter.

The way to understand the mystery is to find a personal balance between the way the asthma limits but pushes at the same time. Not necessarily there yet, I'm trying to move at an appropriate pace, knowing my lungs might just correct me if I falter or rush.

www.ingramcontent.com/pod-product-compliance
Lightning Source LLC
Chambersburg PA
CBHW022019170526
45157CB00003B/1292